Polar Animals

Katie Peters

GRL Consultants,
Diane Craig and Monica Marx,
Certified Literacy Specialists

Pull Ahead READERS

Lerner Publications ◆ Minneapolis

Note from a GRL Consultant
This Pull Ahead leveled book has been carefully designed for beginning readers. A team of guided reading literacy experts has reviewed and leveled the book to ensure readers pull ahead and experience success.

Lerner Publications Company
A division of Lerner Publishing Group, Inc.
241 First Avenue North
Minneapolis, MN 55401 USA

For reading levels and more information, look up this title at www.lernerbooks.com.

Main body text set in Memphis Pro 24/39
Typeface provided by Linotype.

Photo Acknowledgments
The images in this book are used with the permission of: © iStockphoto, pp. 3, 4–5, 6–7, 16 (bottom left, bottom middle); © Shutterstock, pp. 8–9, 10–11, 12–13, 14–15, 16 (top left, top middle, top right, bottom right)

Front cover: © Shutterstock

Library of Congress Cataloging-in-Publication Data

Names: Peters, Katie, author.
Title: Polar animals / Katie Peters.
Description: Minneapolis : Lerner Publications, [2020] | Series: Let's look at animal habitats (Pull ahead readers - Nonfiction) | Includes index. | Audience: Age 4–7. | Audience: K to Grade 3.
Identifiers: LCCN 2018059818 (print) | LCCN 2019000025 (ebook) | ISBN 9781541562042 (eb pdf) | ISBN 9781541558571 (lb : alk. paper) | ISBN 9781541573130 (pb : alk. paper)
Subjects: LCSH: Animals—Polar regions—Juvenile literature.
Classification: LCC QL104 (ebook) | LCC QL104 .P48 2020 (print) | DDC 591.70911/3—dc23

LC record available at https://lccn.loc.gov/2018059818

Manufactured in the United States of America
4-53354-46238-5/11/2022

Contents

Polar Animals

See the polar bears.
They live near the
North Pole.

See the penguins.
They live near the
South Pole.

See the harp seals.
They live near the
North Pole.

See the leopard seals.
They live near the
South Pole.

See the walruses.
They live near the
North Pole.

See the orcas.
They live near the North
and South Poles.

Did You See It?

harp seal

leopard seal

orca

penguin

polar bear

walrus

Index